Published in 2025 by Ruby Tuesday Books Ltd.

Copyright © 2025 Ruby Tuesday Books Ltd.

All rights reserved. No part of this publication may be reproduced in whole or in part, stored in any retrieval system, or transmitted in any form or by any means, electronic, mechanical, photocopying, recording, or otherwise, without written permission from the publisher.

Editor: Mark J. Sachner
Design: Tammy West
Production: John Lingham

Photo Credits:
Alamy: 6 (Colin Varndell), 7R (Arterra Picture Library), 15 (Papilio); Dreamstime: 21 (Cobia); Nature Picture Library: 11 (Paul Hobson), 13 (Dave Bevan), 20R (Guy Edwardes); Ruby Tuesday Books: 12; Science Photo Library: 9 (Eye of Science), 17 (Patrick Landmann); Shutterstock: Cover (Aleksandar Dickov), 5 (Zebra Studio), 7L (Thijs de Graff), 8 (MakroBetz), 10 (Ketta), 14 (Timmary), 16 (Michiel Vaartjes), 18 (Sarah2), 19 (Art Pictures), 20L (Brian Clifford), 22 (Juliana Jus, Nataly Studio, & Waranya Sawasdee), 23 (Eric Isselee, Photoongraphy, & PiortLukasik), 24 (Miroslav Hlavko); Warren Photographic: 4.

ISBN 978-1-78856-438-0

Printed in Poland by L&C Printing Group

www.rubytuesdaybooks.com

CONTENTS

Hello, Little Snail! 4

Glossary 22

Index 24

Hello, Little Snail!

All day, a snail hides in a cool, **damp** place.

Snail

When evening comes, it's time to find food.

A snail has an eye on the end of each long **tentacle**.

Its two short tentacles smell for food.

The snail has a big, stretchy foot made of muscles.

Lots of slime **oozes** out of the snail's body and foot.

The snail moves around on its foot.

Foot

The slime helps the snail stick to plants and walls.

All night, the snail feeds on plants.

It eats with a tongue-like mouthpart called a radula.

The snail's radula has thousands of tiny teeth.

You can only see a snail's teeth through a **microscope**.

Snail's teeth

A microscope picture

The snail eats leaves, fruit and mushrooms.

Mushrooms

It also eats **rotting** leaves and other parts of dead plants.

This helps clean up our world!

Snail slime

A snail's teeth scrape little holes in leaves.

All that eating soon makes the snail need to . . .

. . . POO!

Snail poo

The snail isn't the only animal that's hungry.

Watch out!

The snail quickly hides in its shell so it's not eaten by a mouse.

In the night, the snail meets another snail, and they **mate**.

Each snail is both male and female.

After they mate, both snails lay tiny, white eggs.

Snail eggs

What happens to the snail's eggs?

Eggs

After two weeks, a baby snail will hatch from each egg!

As morning comes, the snail is still eating.

Its slime helps the snail hang upside down on a cabbage.

A slimy slug is also eating the cabbage.

Slug

A slug is a type of snail without a shell.

Today, it feels cold.
It is nearly autumn.

The snail finds a tree hole where it can spend the cold months.

The snail blocks up the opening of its shell with slime.

Hard slime

The slime that blocks the opening gets hard.

See you in spring!

Glossary

damp
Slightly wet. A snail's slimy body needs to stay damp.

mate
To come together to produce young.

microscope
A tool or machine that helps us see things that are too small to see with just our eyes.

ooze
To flow or leak out slowly.

rotting
To break down and get mouldy. When dead plants rot, they become part of the soil.

tentacle
A long body part. A snail can pull its tentacles back into its head.

← Short tentacle inside head

Index

B
baby snails 17

E
eggs 15, 16–17

F
food and eating 4–5, 8, 10–11, 12–13, 18–19

M
mating 14–15

S
shells 5, 13, 19, 21
slime 6–7, 11, 18–19, 21
slugs 19

T
teeth 9, 11
tentacles 5